FIRE & FLOOD

BY

NICOLA BARBER

THE UNCONTROLLABLE ELEMENTS

....

Fires and floods often make headline news. When a fire rages out of control, setting light to everything in its path, or a river bursts its banks sending swirling water through nearby streets and buildings, the results can be disastrous. Although we have developed ways to protect ourselves, we cannot control the power of these natural forces. A fire may start as the result of a lightning strike, or from the heat of lava erupting out of a volcano. A flood is just one part of the great water cycle (see page 10) that distributes water across our planet. Both fires and floods can cause great devastation and loss of life. As we shall see, however, they can also bring great benefits to the environment, wildlife, and to people.

Before the construction of the Aswan Dam on the River Nile in Egypt, the river used to flood once every year irrigating the soil. When the floods went down again, the land was left with a covering of natural fertilizer that helped crops to grow. For centuries, farmers along the Nile relied on its annual floods to raise their crops of barley, wheat, and millet.

LIVING WITH FLOODS

In some regions of the world, people have little choice but to live with regular flooding. The country most severely affected by floods is Bangladesh. This is because about 80 percent of Bangladesh is a vast floodplain for three great rivers, the Ganges, the Brahmaputra (Jamuna), and the Meghna. The country is also often hit by cyclones (hurricanes), which can cause huge storm surges and coastal flooding.

FIRE CLEARANCE

From ancient times, fire has been used to help grow crops. In the tropical rainforests, people cut and burned small clearings in the trees where they planted their crops. When the soil was exhausted they moved on to a new patch, allowing the forest to grow back over the old clearing. This kind of agriculture is known as shifting cultivation, and it is still practiced by some forest peoples.

VITAL FIRE

From the earliest times, fire has been important for survival. Fire provides warmth and can be used to cook food. In many places, fire was an important part of religious ceremonies, and some people worshiped fire — the bringer of both heat and light.

LISBON, 1755

When a huge earthquake hit the city of Lisbon, Portugal, in 1755, it resulted in both flood and fire. Minutes after the quake struck, three huge waves swept up the River Tagus and engulfed much of the lower half of the city. Meanwhile, numerous candles and cooking fires set the wreckage of the devastated buildings alight.

THE VIOLENT EARTH

Deep down beneath your feet, the earth is constantly on the move. Massive splits and collisions in the earth's crust, driven by the molten rock below, are experienced at ground level as earthquakes and volcanic eruptions. Both earthquakes and volcanoes can cause fires and floods. Earthquakes can set off gigantic waves, called "tsunamis," which sweep in to devastate coastal areas (see page 8). Fire is a more indirect consequence of earthquakes, often caused by broken gas mains or upset cooking stoves that set light to the wreckage around them. Volcanoes spew out red-hot lava, which can set fire to vegetation and can spread out over a wide area. Volcanoes can also set off floods of melted ice, ash, and rocks, which form lethal mudslides called "lahars."

SAN FRANCISCO, 1906

The people of San Francisco are used to tremors shaking the ground beneath their feet. But the quake that shook people awake on the morning of April 18, 1906, was more than a tremor. It lasted for minutes and caused buildings to collapse across the city. As the bewildered people fled their homes and started to take in their unfamiliar surroundings, a new threat became apparent — fires started by innumerable upset stoves, broken electric wires, and ruptured gas pipes. The water supplies used to fight fire had been ruptured in the earthquake, and there was little that people could do but watch their city burn for three days and two nights.

ALASKA, 1964

Twisted and broken railway tracks lie along the coast of Alaska, the result of a huge earthquake that hit Alaska on March 27, 1964. The movement of the earth devastated the landscape along the coast of Alaska. Half an hour after the earthquake, the first of several tsunamis hit the coast. These massive waves were over 26 ft. (8 meters) high as they hit the stricken coastline. The tsunamis swept down the coastline as far as northern California.

MUDFLOW MENACE

The city of Seattle, in Washington, is overlooked by Mount Rainier, one of the many volcanoes in the Cascade chain. The city and other towns nearer the volcano are all built on the remains of mudflows that once poured out of the volcano as it erupted. These mudflows are called lahars. They happen when ash and rock from an erupting volcano mix with melting ice from glaciers on the volcano to form a mass of mud that can move at terrifying speed.

FIRE & ICE

When the Icelandic volcano Grimsvötn began to erupt in autumn 1996, scientists across the world became very excited. This was an unusual eruption because Grimsvötn just happens to lie under a huge icecap called Vatnajökull. As the heat of the eruption melted the ice above, everyone waited for the promised deluge. At the beginning of November the water finally broke through, surging for 30 miles (48 km) beneath the ice before emerging from beneath the icecap. The power of the flood left the floodplain along the south coast of Iceland littered with gigantic rocks and chunks of ice 30 ft. (9 meters) high. This type of flood is called a jökulhlaup, an Icelandic word meaning "glacier flood."

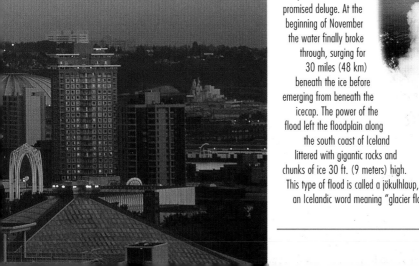

A spectacular storm sends streaks of lightning to the ground in Tucson, Arizona. Not all lightning strikes start fires, but lightning fires are much more common in some parts of the world than others. Scientists say that lightning starts about 65 percent of all wildfires in western United States, but only about eight percent of bushfires in Australia.

FIRES: WHY DO THEY HAPPEN?

T he two main causes of fire in the environment are **lightning and human activity.** On average, lightning strikes the ground 100,000 times every day — although not all these strikes start a fire. However, it is people who cause the majority of countryside fires, mostly through carelessness. Throwing away a lit match or glowing cigarette butt can have catastrophic consequences, but a surprising number of fires are started deliberately. In some places, a fire can start without any immediately obvious cause. This is called spontaneous combustion, and it can happen when there is a large buildup of dead and rotting vegetation, which produces heat.

BLACK DRAGON FIRE

In the northeast of China lies the Hinggan forest. Through the middle of the forest runs the Black Dragon (Amur) River. For centuries, lightning fires have burned through the forest, leaving scars that are eventually covered over with new growth. But the fire that broke out on May 6, 1987, started from lit cigarettes igniting some spilled petrol. It was quickly spread by strong winds, and despite attempts to control it, the fire burned until May 12. It destroyed about 20 percent of China's forest reserves, killed over 200 people, and left over 50,000 homeless.

A campfire should be set on rock or bare ground to prevent vegetation around the fire from igniting. In some wilderness areas fires may not be permitted because of the high risk of forest fires.

STARTING A FIRE

Most fires in the countryside are started by carelessness.
Many people do not realize the danger of throwing away a lit cigarette or match into dry undergrowth, or of not controlling and putting out a campfire properly. Another cause of fire is the sparks from machinery — from wheels of trains as they run along metal tracks, or from equipment used by lumber companies. Meanwhile, it is estimated that about 25 percent of all forest fires in the United States are started on purpose.

FIRE TRIANGLE

What do you need to make fire? The three essential ingredients for a fire are heat, fuel, and oxygen. These make up the "fire triangle." For a fire to start, the fuel must heat to a certain temperature (called the ignition point) and there must be oxygen for the fuel to react with. If one part of the fire triangle is missing, a fire cannot burn. For example, if you throw a fire blanket over a small fire, the fire should go out because it is starved of oxygen.

SEASONAL FLOODS

Flooding doesn't stop life from carrying on as normal in Varanasi, northeast India. Varanasi stands on the River Ganges, and the city streets are often filled with water during the summer monsoon season. There are two monsoon seasons in southern Asia: the winter monsoon brings hot, dry weather; the summer monsoon brings heavy rains, which swell rivers and cause regular flooding.

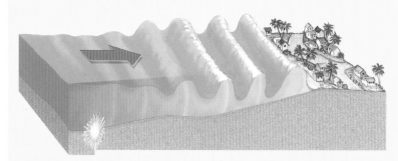

LYNMOUTH, 1952

The town of Lynmouth, in Devon, England stands at the mouths of two rivers, the West and East Lyn, which tumble 1,476 ft. (450 meters) down picturesque gorges before reaching the sea. The rainfall on August 15, 1952, was one of the heaviest ever recorded in Great Britain, and both rivers were soon swollen with water. During the night, a wall of water over 30 ft. (9 meters) high swept through the town, demolishing houses in its path, sweeping cars far out to sea, and depositing huge boulders on the beach. This flash flood killed at least 34 people.

TSUNAMIS

This artwork shows a tsunami about to swamp a coastal village as a result of an earthquake under the ocean bed. In the middle of the ocean the tsunami is fast-moving but often barely detectable, with a crest only about 3 ft. (1 meter) high. But, as it reaches shallower water, the tsunami slows down and the water in the wave starts to pile up. The biggest tsunamis can reach up to a terrifying 82 ft. (25 meters) in height.

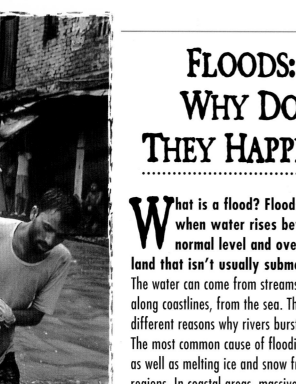

FLOODS: WHY DO THEY HAPPEN?

What is a flood? Floods happen when water rises beyond its normal level and overflows onto land that isn't usually submerged.
The water can come from streams and rivers or, along coastlines, from the sea. There are many different reasons why rivers burst their banks. The most common cause of flooding is heavy rain, as well as melting ice and snow from mountainous regions. In coastal areas, massive storms such as hurricanes can whip up huge waves, pushing vast amounts of water inland. Floods can also result from landslides, dam failures, earthquakes, and volcanic activity. Often it is a combination of several problems that leads to flooding. Flash floods are particularly dangerous. These violent floods happen suddenly and often without any warning.

A CATASTROPHIC FAILURE

Dam failure, when a dam breaks and allows water to escape, has been the cause of some major floods. But, during the catastrophe that overwhelmed an Alpine valley in 1963, the dam did not fail. The Vaiont Dam was built 846 ft. (258 meters) high at the head of a valley in northeastern Italy. Above it stood the steep slopes of Mount Toc. On the night of October 9, after heavy rain, the unstable slopes of the mountain slid into the reservoir, sending a massive wave 325 ft. (99 meters) over the top of the dam. In only 15 minutes, this giant wave swept through the valley below the dam, killing 2,600 people.

STORM SURGES

Dutch troops haul sandbags to try to hold back the waters along the low-lying Netherlands coastline in 1953. In January of that year the coasts of the Netherlands and eastern England were battered by storms. Toward the end of the month, the gales coincided with a very high tide and, during the night of January 31, the seas poured over the tops of the dykes protecting land and homes in the Netherlands. The storm surge flooded over 247,000 acres (100,000 hectares) of Dutch farmland with salt water, and drowned over 1,500 people.

SIGNS & WARNINGS

If lives are to be saved, it is vital to have good systems to warn people where and when fires and floods are likely to happen. Weather forecasting is the first priority. Today, meteorologists (weather scientists) can predict quite accurately what the weather will do for a few days ahead. Fire researchers use this information together with their specialized knowledge to assess the likelihood of a fire. If a fire is already burning, they look at factors such as temperature, wind direction, and speed to try to work out how the fire will behave. Hydrologists (scientists who specialize in the behavior of water) also use weather information. However, predicting exactly where a flood might happen isn't easy. While heavy rain in one area may run away harmlessly, a cloudburst in another area may cause a flash flood. Hydrologists use their knowledge about each particular area to assess the risk of flooding.

HURRICANE ALERT

This satellite picture of Hurricane Andrew shows the distinctive whorl shape of hurricane clouds with the "eye" at the center. Big hurricanes bring torrential rain and strong winds. If they hit coastal areas they can cause huge waves and storm surges. Using satellite pictures, meteorologists track hurricanes and try to predict where they will go next. If a severe hurricane threatens to hit a populated area, whole regions are sometimes evacuated.

water falls back to the ground in the form of rain and snow

trees and plants give off water vapor

water vapor from oceans, lakes, and rivers evaporates and cools to form clouds

some water seeps into the soil until it becomes trapped by rock; it flows down, back into the rivers and oceans

THE WATER CYCLE

All the water on Earth is involved in an endless cycle in which it moves from the surface of Earth into the atmosphere and then back again. This constant movement is called the water (or hydrological) cycle. Water rises into the atmosphere, cools, condenses, and returns to Earth's surface as rain or snow. Then the water either evaporates straight back into the atmosphere, is carried by rivers to the sea, or goes deep underground. Some is locked as ice into icecaps and glaciers. It may take thousands of years, but eventually all of this water will evaporate back into the atmosphere, where it will once again condense and fall back to Earth.... And so the cycle continues.

RAIN DATA

This type of rain gauge is used to measure the amount of rainfall on the ground. Simple rain gauges have a collecting funnel and a dipstick that shows the level of water. More complicated devices are designed so that hydrologists can telephone the number of the gauge to get the latest reading. Other gauges use radio waves, microwaves, or communications satellites to transmit their data.

In the U.S., a flood watch/flash flood watch is issued when a flood is possible. Flood warnings/flash flood warnings are more serious and mean that floods are expected in your area and you should evacuate to higher ground immediately.

FLOOD WARNING

The United States has a very well-organized and up-to-date system of flood warning run by the National Weather Service of National Oceanic and Atmospheric Administration (NOAA). Information comes from the National Meteorological Center and is fed to 13 River Forecast Centers. Hydrologists at these centers prepare a flood forecast that is passed on to Weather Forecast Offices. If necessary, warning of a flood is issued to radio, television, newspapers, police, and other agencies. Many other countries operate similar systems.

A fire alert board in the United States warns of the fire danger in a particular area. The board shows that this region is on the second highest alert for fire. The risk of fire is assessed by collecting information from various sources about the weather (temperature, wind, and rainfall) and the amount of "fuel" lying on the ground in a particular area and how dry it is.

CLIMATE CHANGE

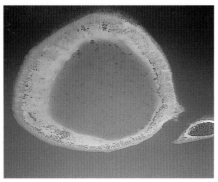

RISING SEAS

Some scientists predict the sea level could rise enough to flood many coastal areas, and to leave some islands, such as the Maldive Islands, permanently underwater. Global warming could cause the vast ice sheets at the North and South Poles to melt, unlocking huge amounts of water into the seas and oceans.

1998 was a bad year for fires and floods. Heatwaves caused droughts in many parts of the world leading to massive forest fires in Southeast Asia and the Amazon. In Florida, thousands of people were evacuated from their homes because of forest fires. Meanwhile, torrential rain brought terrible flooding to Bangladesh and to China, where over 2,500 people drowned. Finally, Hurricane Mitch devastated Honduras, killing more than 10,000 people. All of these events made 1998 the worst year on record for natural disasters, and scientists are predicting that this is the pattern for years to come. Many think that one of the reasons for the increase in flood and fire disasters in the 20th century is global warming. As the world heats up, there is more energy to drive the water cycle (see page 10) and other climate systems resulting in more extreme temperatures, rainfall, and winds.

MITCH WREAKS HAVOC

In late September 1998 a hurricane formed over the South Caribbean Sea. It was named Hurricane Mitch, and it was destined to be the most ferocious hurricane to hit the Caribbean since Hurricane Gilbert in 1988. At the end of October, Hurricane Mitch swept across Honduras in Central America, causing catastrophic flooding and mudslides. By November 2, over 10,000 people had died in Honduras.

BEING PREPARED

The summer of 1993–94 was extremely hot and dry in Australia. There was no rain for weeks, and temperatures were regularly above average. All of this pointed to an extremely high fire risk and, in January 1994, several bushfires broke out. The fires swept through New South Wales and threatened the outskirts of Sydney. Yet, good organization and preparation meant that total disaster was averted and, amazingly, only four people lost their lives.

VOLCANIC POLLUTION

The gases causing global warming are known as "greenhouse gases" and come from air pollution from industry, vehicle exhausts, and from nature itself. Erupting volcanoes spew out huge clouds of ash and gases that can affect weather worldwide. The haze that floated in the air after a huge eruption in Iceland in 1783 reduced the amount of sunlight reaching Earth's surface. The result was famine in Iceland and some of the coldest winters on record. More recently, the ash from the 1991 eruption of Mount Pinatubo in the Philippines also affected weather patterns around the world.

IN THE GREENHOUSE

Why is the world heating up? The most likely explanation seems to be that temperatures are increasing as a result of the greenhouse effect. Heat and light from the sun are vital for life on Earth. But an increase in the levels of some gases in Earth's atmosphere means that more heat is becoming trapped — just like heat energy is trapped inside a greenhouse. More heat energy means higher temperatures.

LIVING WITH THE RISK

If there is going to be more extreme weather in the future, as many scientists predict, this is bad news for the millions of people who live in flood-risk areas. Bangladesh is one of the poorest and most densely populated countries in the world. Over 80 percent of Bangladeshis live in the countryside and rely on agriculture for a living. Land is essential for their survival, yet the country is at regular risk from cyclones, storm surges, and disastrous river floods.

SOUTHERN FRANCE

Southern France has a high fire risk during its hot, dry summer. A dry wind known as the "mistral" blows regularly from the north, funneling down valleys at high speeds and spreading any fire in its path. Every year there are between 2,000 and 3,000 fires in the region — mostly caused by human carelessness.

In 1997, over 20 million people across Indonesia, Malaysia, and Singapore were treated for smoke-related illnesses, such as asthma or bronchitis, as a result of forest clearances. Cotton face masks were scant protection against the acrid haze.

SOUTHEAST ASIA

In October 1997, the streets of Jambi in Sumatra were filled with smoke from fires, mostly started by big businesses intent on clearing the forest to prepare land for planting pulp wood and oil palm plantations. Although fire clearance is officially forbidden in most areas, the law is widely ignored because fire is the quickest and cheapest method to clear large areas of land.

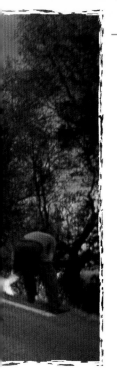

HOT SPOTS

One of the world's hottest spots in 1997 was Southeast Asia. Fires in Indonesia spread a huge cloud of smoke that hung as a thick haze across an area larger than Europe. The worst fires burned in the tropical forests of Sumatra and Kalimantan (the Indonesian part of Borneo). But the smoke affected people across Indonesia, Malaysia, Brunei, and Singapore. Despite predictions of drought, fires started in May and June, and by September they were raging out of control in many places. In the smoke haze from the fires, visibility was so bad that ships ran aground and collided. The smoke even caused a passenger airplane to crash in Sumatra, killing over 200 people. Only when the seasonal rains came in November did the fires finally damp down and the smoke disperse.

WESTERN UNITED STATES

In 1988 a prolonged drought parched much of the western part of the United States. Following a drought in the previous year, and lower-than-average snowfalls, the fire risk was critical by May of 1988. The first fire in Yellowstone National Park was started by lightning on May 24 and was quickly put out by rain. However, other longer-lasting fires soon followed. By the end of 1988, fire had scorched nearly 800,000 acres (320,000 hectares) of the national park.

SOUTHEAST AUSTRALIA

Eucalyptus trees catch alight very easily because their bark contains highly combustible oils. They also produce lots of litter, which provides excellent dry fuel for a fire. Southeast Australia is one of the most fire-prone regions in the world. During the long summers, hot, dry winds blow in from the desert interior of the continent. Even during the winter, there is usually little rainfall. This means that the environment is usually dry and that fires start and spread very easily.

BURNING FOREST

Amazonia is the largest area of rainforest on Earth, but just as in Indonesia, large companies have been burning vast areas of rainforest as a quick solution to clear areas of land for plantations. Satellite images of South America help scientists to monitor just how fast the rainforest is disappearing. For example, in 1987 it is estimated that over 77,220 square miles (200,000 square km) of rainforest were destroyed by fire.

FIGHTING A WILDFIRE

Fires can cause huge amounts of damage to property and the environment. They can also threaten human life. These threats mean that, in many parts of the world, governments spend many millions of dollars on fire detection and firefighting. Spotting a fire quickly, before it has a chance to take hold and spread, is a priority. Hi-tech equipment, such as infrared scanners mounted in planes, help to pinpoint fires. But in many places the best way to spot a fire is still through human eyes from a fire lookout post. Once the fire has been spotted, firefighters can go into action. Water and foam are sprayed by helicopters and airplanes. Strips of land are bulldozed to act as firebreaks. Planes also drop parachutists, called "smokejumpers," into isolated areas to fight fires.

MAKING A BREAK

A firebreak is a wide corridor that is opened up in a forest. By removing all the trees and other material from the firebreak area, the fire is starved of fuel. Since firebreaks make big scars across the landscape, they are used mainly for big fires.

Firefighters in California spray water from the ground and from the air to try to prevent a fire from jumping across a road. Roads, railway tracks, rivers and lakes, and bare rock ledges all act as natural firebreaks. Spraying water on the vegetation on either side of the firebreak makes it less likely that the fire will be able to take hold.

This plane is dumping red retardant onto a forest fire. Retardant is used to slow the advance of a forest fire.

DAMPING DOWN

To put out a fire, firefighters need to remove one of the three essentials for fire: heat, oxygen, or fuel. Dumping water onto a fire helps to lower the heat of the fuel. Planes and helicopters are also used — some have tanks that are pumped full of water before takeoff, others are equipped with scoops so that they can fly over the sea or a lake and refill their tanks with water.

SMOKEJUMPERS

Sometimes the only way in to fight a fire in a remote wilderness area is by air. Firefighters called smokejumpers are dropped in by parachute. Their job is extremely hazardous and requires rigorous training. They have to carry firefighting and survival gear, as well as their parachutes. Often, the firefighters must walk out of the wilderness once the fire is extinguished carrying a backpack weighing up to 110 lbs. (50 kg).

AFTER A FIRE

Fire has been a regular part of life for plants and animals over millions of years. Many plants have adapted to fire, with barks or leaves that can resist the effects of heat. Others actually rely on fire to make their seeds germinate. Fire can increase the fertility of the soil by releasing nutrients (such as potassium, magnesium, calcium, and phosphorus) from the burned vegetation. However, an intense fire can also destroy nutrients in the soil. Faced with fire, animals show little sign of panic but are adept at finding refuge — although a fast-spreading fire will inevitably overwhelm and kill some animals. The biggest threat for most animals is a change in habitat. If an animal relies on only one kind of food, which is destroyed by fire, then that animal may not survive.

TOUGH SEEDS

Some plants, like this Protea, produce seeds with thick outer casings that need the heat of a fire to open them. These seeds can sit in the soil for many years. Only when the outer casing cracks can the seed inside start to germinate and grow.

SEQUOIA SECRETS

Giant sequoias depend on fire to help them flourish. Without fire, other shrubs tend to shade sequoia seedlings and prevent them from growing. A fire will burn competing shrubs, giving the sequoia trees more space and light. The sequoia is well adapted to survive a fire. It has deep roots to avoid damage from a ground fire. Its thick bark protects the living part of the massive trunk from burning, and it has high branches that can escape most fires.

FIRE PINES

Some plants need fire in order to release and germinate their seeds. Many types of pine tree, known as "fire pines," produce seeds that are contained inside closed cones. The cones are held shut by a sticky substance called resin. During a fire, the mature tree may be killed but the cones fall to the ground and the resin melts releasing thousands of seeds onto the newly burned and fertilized soil. "Fire pines" can be found in many countries where there is a risk of fire, like this Lodgepole pine in North America.

BORNEO, 1983
MONKEY SURVIVAL

After massive fires in the rainforests of Borneo in 1983, scientists studied how certain animals adapted to the changed habitat. They noted that a group of macaques quickly changed their pre-fire diet of fruits, seeds, and flowers to dried fruits, plants, and insects after the fire. Many other small animals make similar adaptations, allowing them to survive in a changed environment after a fire.

YELLOWSTONE SURVIVOR

There are about 30,000 elk in Yellowstone National Park, and about 240 animals perished during the fires of 1988. Most died from the effects of smoke as the fires were swept through the park by strong winds. Many animals, such as bears and bald eagles, came into the park right after the fires to feed on the carcasses of animals killed in the fire.

PREVENTION & SCIENCE: FIRE

The best way to stop a fire is to never let it start. As we have seen, most environmental fires are started by people, either through carelessness or accidents. Making people more aware of the danger of fire is a big step toward reducing the number of fires; educating people on how to deal with fire once it strikes is equally important. Firefighters spray houses with foam to try to protect them from the advance of a forest fire; foams are also used directly to fight wildfires. They are nontoxic and break down afterwards without harming the environment. Scientists use satellites with infrared (IR) sensors to detect heat, and light sensors to record smoke and other signs of fire. However, clouds block these sensors making them an unreliable source of information. Scientists continue to work on satellite technology for fire detection, but more traditional fire alert systems are more reliable at the present.

EAGLE-EYE VIEW
Fire lookout posts are the main method of fire detection in many places. Some lookout towers are equipped with TV cameras that scan the landscape below, sending a picture back to a central base. However, it is often difficult to tell the difference between smoke and mist on a TV screen, making the camera lens a less effective fire detector than the human eye. Fire lookout posts are often staffed by volunteers who report any fires to a central control by radio. This one in Oregon is no longer used as a lookout post but has been renovated for use by campers.

FIGHTING FIRE WITH FIRE

Fire can be used itself as a weapon to fight fire. Firefighters sometimes set fire to the vegetation that lies in the path of an approaching fire. By controlled burning of a small area they can create a firebreak that starves the main fire of fuel. Fire is also used to prevent possible future fires. This is known as prescribed fire. It is used as part of land management schemes, often to keep fuel levels in check.

SELF-PROTECTION

If you live in a fire-prone area, or you are faced with a wildfire, there is a lot you can do to help yourself. Here are just some of the guidelines issued to people who live in fire-risk areas in Australia.

○ Fuel reduction is vital: ensure that fallen leaves, long grass, and dead undergrowth is regularly cleared away. Clear a firebreak at least 7 ft. (2 meters) wide all around your property.

○ Make sure you have a good water supply without having to rely on water mains. Keep a supply of water in a tank or in a swimming pool.

○ Check your equipment, particularly ladders and hoses. Make sure your hose can reach all sides of the house.

○ Prepare for your own personal safety. You will need long-sleeved shirts and long pants made from natural fibers (e.g., cotton or wool: man-made fibers such as polyester or nylon can melt and burn the skin), sturdy leather boots, gloves, goggles, and a broad-rimmed hat.

WET SPOTS

One of the wettest spots on Earth in 1993 was along the Mississippi River. After a wet spring, thunderstorms brought more rain to the Midwest in the summer — and they stayed. For months the rain fell, making it the wettest summer on record in states such as Minnesota, Illinois, and Iowa. Between April and July, Iowa was inundated with as much rain as it usually gets in a whole year. The rainfall swelled the Mississippi River to record levels. The floods started in Minnesota in June, and through the summer they progressed downstream. Towns and farmland along the Mississippi are protected by huge earth banks, called levees, which are designed to contain the river even during a flood (see page 24). But, along the length of the Mississippi, the water cascaded over levees more than 20 ft. (6 meters) high. Damage to property and farmland was estimated at 10 billion dollars.

MIGHTY MISSISSIPPI

Floods are a frequent occurrence along the length of the Mississippi but the floods of 1993 were particularly devastating. Many two-story houses were flooded to the rooftops. The Mississippi is 2,405 miles (3,780 km) long. It runs from Lake Itasca, Minnesota, to the Gulf of Mexico and, together with its tributaries, it drains an area of about 1.2 million square miles (3.2 million square km) including 31 U.S. states and two Canadian provinces.

FLOODED CITY

The beautiful city of Venice in Italy is one of the most unusual in the world. It is built on about 120 small islands in the middle of a lagoon at the north end of the Adriatic Sea. Instead of streets it has canals. Not surprisingly, parts of the city are flooded every year by high tides and storms.

CHINA'S SORROW

The Huang Ho (Yellow River) flows 3,380 miles (5,440 km) from the Himalayas across a vast plain in northern China to the East China Sea. It is known as "China's Sorrow" because, over the centuries, its floods have killed millions of people. For centuries the Chinese have struggled to contain the river by building banks, called dykes, from willow branches, kaoliang (a type of sorghum) stalks, stones, sand, and bricks.

THE LOWLANDS

For over 700 years, the Dutch people have built dykes and dams to stop storm surges from flooding their low-lying, marshy coastline. Much of the farmland that lies along the Dutch coast today is land that has been reclaimed from the sea, called polders. This reclaimed land is either at sea level or below it, and needs to be protected by dykes.

FLOODS IN THE DEVELOPING WORLD

People in Dhaka, Bangladesh, travel on a wooden boat called a nouka, their umbrellas providing scant protection against the monsoon rains. Floods are the most frequent natural disaster to affect the developing countries of the world. These countries do not have the money to spend on expensive flood protection schemes and many of their floodplains and coastal areas are densely populated. This means that floods and cyclones often have very severe impacts when they strike.

FLOATING HOUSES

The people in the town of Iquitos, Peru, live alongside one of the greatest rivers in the world – the Amazon. Iquitos lies in the shadow of the Andes Mountains, and this part of the river is prone to flooding. But the town of Iquitos rides up and down with the floods. The cane and thatch houses are built on log rafts that sit in a shallow bend of the river.

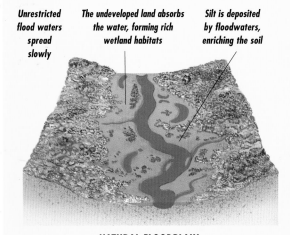

GROWING RICE

Farmers in Bangladesh make the most of the flooding that is part of everyday life in the country. Rice needs water to grow, and it flourishes on the floodplains and deltas of rivers. The seasonal floods in Bangladesh mean that farmers can grow three crops of rice a year. However, severe or unexpected floods can devastate crops, uprooting and damaging plants, and dumping huge amounts of silt in paddy fields.

BUILDING LEVEES

As long as people have lived near rivers, they have built embankments to try to protect themselves from flooding. In the United States, these embankments are known as levees. The first levee was built along the Mississippi in 1718 to protect the settlement of New Orleans. Since then, the struggle to contain the Mississippi has continued, but levees bring their own problems. They may stop water overflowing in one place, but the extra water can cause worse flooding elsewhere.

Unrestricted flood waters spread slowly

The undeveloped land absorbs the water, forming rich wetland habitats

Silt is deposited by floodwaters, enriching the soil

Water can push over levees, ruining less protected crops

Levee walls, built to protect towns and crops, can cause floodwaters to swell upstream

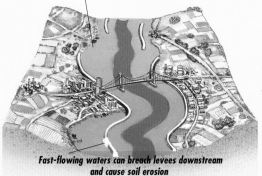

Fast-flowing waters can breach levees downstream and cause soil erosion

NATURAL FLOODPLAIN

LEVEE-RESTRICTED WATER

LIVING WITH THE THREAT

Some people have always lived with the threat of floods. This is because some of the best agricultural land in the world is found on floodplains. Just like in Ancient Egypt, farmers around the world have long relied on the nutrients deposited by river floods to fertilize the soil. Rivers are also vital links for transportation and trade. Over the centuries, as people have settled near rivers, they have also tried to protect themselves from the threat of flood.

TSUNAMI PROTECTION

Huge concrete slabs form a coastal defense in Japan. Japan has experienced many destructive tsunamis in its history. This type of sea wall helps to absorb the energy of the wave that might otherwise sweep in to destroy buildings.

As long as 2,500 years ago, people in China were building mud banks to try to control the unruly Huang Ho. In Britain, the Romans built floodwalls to protect parts of the coastline. The building of flood barriers, dams, and reservoirs continues to this day.

MOVING HOUSE

Some people who live near the Mississippi have had enough of coping with the river's floods. After the floods in 1993, many communities decided to make a move to safer and higher ground. However, this kind of solution is only available in rich countries, where there is enough wealth to pay for settling in a new area, and where the population density is not too high — so that there is free land to move to.

AFTER A FLOOD

What happens after the floodwaters have gone down, and people start to return to their shattered homes and land? The first and most devastating consequence of flooding is loss of life. Bodies can be swept far away in a flood; some may never be found. People with injuries need to be evacuated and cared for. Then the cleanup operation can begin. The power of floodwaters often smashes houses and washes away foundations. Floodwaters also often carry vast amounts of stinking mud that can cover everything in sight once the waters have gone down. Another problem is that floodwaters are not clean — the floods run into sewers, spreading sewage through houses and streets and increasing the risk of diseases such as cholera, malaria, and dysentery.

A woman cleans up after the floods that devastated much of the coast of the Netherlands in 1953. The horrible mess left by flood waters is clear in this picture. Over 70,000 people were evacuated from the area during the floods, and over 400,000 buildings were damaged by the force of the waters.

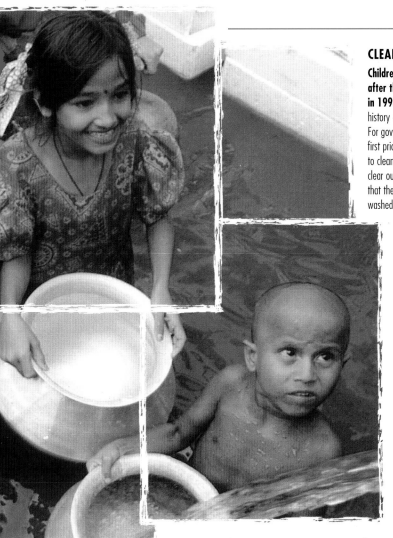

CLEAN WATER

Children collect drinking water from a truck after the floods that hit Dhakar, Bangladesh, in 1998. These floods were the worst in the history of Bangladesh and left millions homeless. For governments and aid agencies, one of the first priorities is to ensure that people have access to clean water supplies. The next priority is to clear out and clean up wells, and make sure that they are not contaminated by sewage washed in by the floodwaters.

HONDURAS, 1988

The capital of Honduras, Tegucigalpa, was devastated by floods caused by Hurricane Mitch as up to 4 inches (10 cm) of rain fell per hour at the height of the storm. An eyewitness described her helplessness as..."several neighbors' homes [were] washed away in the storm. The house below ours filled with mud to the eaves. We moved out of our house...when the bank below us began to crumble away...Many of those who rescued others had themselves watched their own homes flooded or jerked away by the currents..."

RUINED CROPS

Farmers stand in the ruins of their fields in Honduras surveying the devastation caused by Hurricane Mitch in 1998. The hurricane winds shredded vegetation, and torrential rain caused massive mudslides and floods. Over half of all the crops in Honduras were destroyed by the storm. Aid agencies provided food relief for over half a million people who were in danger of starvation. Another priority was to distribute seeds, so that fields could be cleared and replanted in good time to provide crops for the next season.

ART RESCUE

"The Arno is out!" This was the cry around Florence, Italy, on the night of November 4, 1966. After very heavy rain, the River Arno finally flooded the historic city. The floods rose to 20 ft. (6 meters), causing extensive damage to many important historical buildings, artwork, and sculptures. One of the main problems was the oil that leaked out of central heating units and coated everything it touched. In the library, more than a million books were covered in thick, oily mud, some dating from Renaissance times. Volunteers were forced to wear gas masks to guard against the stink of sewage and rotting leather. Experts searched the mud for tiny fragments of damaged frescoes, and cleaned and restored paintings. The huge international effort helped to rescue many priceless works of art.

PREVENTION & SCIENCE: FLOODS

What precautions can be taken against floods and what scientific advances give hope for the future? All around the world, meteorologists and hydrologists make use of the latest hi-tech equipment to monitor water flows and rainfall. Scientists can now obtain more accurate information from rain gauges mounted in aircraft or on satellites. Weather radar is also used and, combined with information from ground gauges, readings can be given for whole areas. All this information is processed by powerful computers. To control rivers and protect their coastlines, the wealthy, developed nations often have extensive and complicated systems of dams, reservoirs, dykes, and barrages — although even the most hi-tech schemes may fail in the event of a "super flood."

WATCHING FOR WAVES

After the tsunami that hit Hawaii in 1946, it was decided to set up a warning system in the Pacific region. This system is now known as the Tsunami Warning System, and 26 nations work together to collect data and issue warnings. This system makes use of satellites and sensors all across the Pacific region that alert weather offices if a tsunami is on its way. However, warnings of tsunamis sometimes do not have the desired effect. After a warning of a tsunami in 1964, thousands of people flooded onto the beaches of San Francisco to watch!

VENICE IN PERIL

In 1996, St. Mark's Square in Venice was flooded for 101 days. This was the worst year on record. Venice is threatened by two main problems: rising sea levels as a result of global warming, and the fact that Venice itself is sinking. For years, scientists and politicians have been arguing about the best way to protect Venice. One plan is called the Moses project. It would involve fixing 79 hinged barriers to the seabed at the mouth of the lagoon. These barriers would be raised in the event of a higher than normal tide. However, many people are concerned about the effect of a barrier on wildlife in the lagoon. Another problem with a barrier, as with all coastal defenses, is that the water it deflects could cause worse flooding further along the coast.

HOW THE BARRIERS WOULD WORK?

The barriers would remain on the seabed until the tide is 3 ft. higher than normal.

WATER

To raise the barriers, compressed air would pump out the water and the barriers would float to the surface.

LAGOON   AIR   SEA

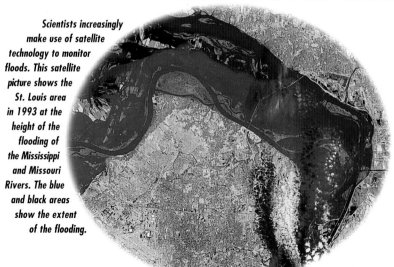

Scientists increasingly make use of satellite technology to monitor floods. This satellite picture shows the St. Louis area in 1993 at the height of the flooding of the Mississippi and Missouri Rivers. The blue and black areas show the extent of the flooding.

HI-TECH BARRIERS

Many countries have invested in hi-tech solutions to protect important places from flooding. Since 1983, London has been protected from a storm surge such as the one that hit eastern England in 1953 by the Thames Barrier. Normally the barrier lies flat on the riverbed. But if warning of a storm surge is received, the gates are raised so that they stand 52 ft. (16 meters) above the riverbed. It takes about 30 minutes to raise the gates completely.

FLOOD MYTHS

Myths about floods are found in cultures all over the world.

In these stories, a flood is often sent by the gods to punish wicked people on Earth. In the Bible story of Noah, the Ark contains two of every bird and animal as well as Noah's family. The Ancient Babylonians told the story of Enlil who sent a great flood to quiet the peoples of the world. The Hindu story of Manu relates that Manu was warned by a fish to build a ship. When the floods came, the fish pulled the ship and Manu to safety.

The Aztec people of Central America believed that, in the past, the world had been destroyed by both fire and flood. This statue is of the Aztec rain god Tlaloc. Both the Aztecs and the Incas of South America also worshiped fire gods.

UNCOVERING THE PAST

There have been fires and floods as long as humans have been living on Earth. Ancient peoples told powerful stories, called myths, to try to explain these terrifying natural phenomena. Scientists today know about some ancient fires and floods from archaeological deposits and remains. A layer of clay and debris deep in the ground can indicate a flood thousands of years ago. A layer of charcoal can indicate an ancient fire and can be dated quite precisely. Trees can also reveal useful information. Both floods and fires can leave scars on a tree, which show up in the tree's growth rings. These scars can be used to date floods and fires, a technique known as dendrochronology.

THE PHOENIX

The story of the Phoenix comes from Greek mythology. Some writers relate that this brightly colored bird lived for 500 years, some that it lived for up to 12,954 years. At the time of its death, it burned itself on a pyre of spice tree twigs. Then a new Phoenix rose from the ashes with renewed youth and beauty. The young Phoenix took the ashes of its father to the city of Heliopolis in Egypt, the City of the Sun.

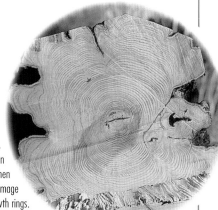

DENDROCHRONOLOGY

The growth rings in a tree may tell the story of a fire or flood. In the year of a fire the ring may be narrower, but in following years the rings are often wider than average because of the added nutrients in the soil as result of the fire. A flood can damage a tree when debris carried in the waters crashes into a tree, damaging the living wood beneath the bark. This damage shows up in the growth rings.

ANCIENT FLOOD

Some of the earliest-known civilizations on Earth were on the floodplains of the Tigris and Euphrates Rivers in an area known as Mesopotamia (modern-day Iraq). The peoples of this area must have experienced many floods. In 1929, excavations at the ancient city of Ur uncovered a thick layer of clay deep under the ground. This 10 ft. (3 meter) layer indicated that many thousands of years ago there had been a massive flood, probably up to 25 ft. (7.5 meters) deep.

IRAQ, 1929
NOAH'S FLOOD

The archaeologist in charge of the excavation at the ancient city of Ur was a British man, Sir Leonard Woolley. After examining the thick layer of clay flood deposits he wrote, "We had thus found the flood on which is based the story of Noah." For the ancient people caught in this flood, it must have seemed as if the deluge covered the whole Earth.

DID YOU KNOW?

About 98 percent of the water on the Earth's surface is in the Pacific, Atlantic, and Indian Oceans.

In the open sea, tsunamis can travel at up to 497 m.p.h. (800 km/h).

After the eruption of Krakatoa in 1883, the tsunamis that hit Java and Sumatra were so huge that large shock waves were detected as far away as the English Channel.

The 16th-century artist and inventor, Leonardo da Vinci, drew up plans for flood defenses to control the River Arno and to prevent Florence from flooding. But...his plans were rejected.

In 1931, both the Huang Ho and the Yangtze Rivers broke their banks after a heavy rain. The Yangtze rose to 95 ft. (29 meters) higher than its normal level! The flooding caused the deaths of millions of people.

In 1938, the Chinese deliberately broke the dykes of the Huang Ho in order to stop a Japanese invasion.

The flooding halted the Japanese attack — but it also drowned over 500,000 Chinese people and left many more homeless.

From archaeological evidence, we know that early humans were using fire as long as 60,000 years ago. They learned that the fresh green grass that grew on burned soil attracted large herds, so they began to burn areas of grassland on purpose.

During the drought of 1988 in the United States, there were over 70,000 wildfires that burned over 5 million acres of land.

There is fuel for a forest fire at surface level, below ground, and in the air. A ground fire burns material on and below the forest floor. A surface fire spreads along grasses, shrubs, and small trees as well as the layer of freshly fallen leaves, logs, and branches. A crown fire spreads from treetop to treetop. This is often the fastest spreading fire, leaping from one tree to the next faster than a person can run.

ACKNOWLEDGMENTS

We would like to thank: Graham Rich, Hazel Poole, and Elizabeth Wiggans for their assistance. Artwork by Peter Bull Art Studio.
First edition for the United States, Canada, and the Philippine Republic published by Barron's Educational Series, Inc., 1999.
First published in Great Britain in 1999 by ticktock Publishing Ltd., The Office, The Square, Hadlow, Kent TN11 0DD, United Kingdom.
Copyright © 1999 ticktock Publishing Ltd. American edition Copyright © 1999 Barron's Educational Series, Inc.
All rights reserved. No part of this book may be reproduced in any form, by photostat, microfilm, xerography, or any other means, or incorporated into any information retrieval system, electronic or mechanical, without the written permission of the copyright owner.

All inquiries should be addressed to: Barron's Educational Series, Inc., 250 Wireless Boulevard, Hauppauge, New York 11788
http://www.barronseduc.com

Library of Congress Catalog Card No.99-62338
International Standard Book Number 0-7641-1058-6

9 8 7 6 5 4 3 2 1   Picture research by Image Select. Printed in Hong Kong.

Picture Credits: t = top, b = bottom, c = center, l = left, r=right, OFC = outside front cover, OBC = outside back cover, IFC = inside front cover

Colorific: 3tr, 6cr, 11br, 12br, 15tr, 17tr, 21t, 22tl, 22c, 23tl, 22/23 (main pic), 24cl, 25b, 24/25t, 28c. Corbis: 4tl, 5tr, 17cr, 18/19 (main pic), 20/21 (main pic), 25tr, 27cb, 30/31 (main pic). e.t.archive: 28tl. Mary Evans Picture Library: 31tr. National Geographic: 8/9t, 22bl. Oxford Scientific Films: OFC (main pic), OFC (inset pic), IFC, 18tl. Planet Earth: 12tl, 12/13 (main pic), 14/15b, 16tl, 18bl, 18/19b. Popperfoto: 2bl, 5br, 8cl, 9br, 9tr, 14bl & OBC, 16bl, 26cl, 26/27 (main pic). Rex Features: 14/15t, 16/17 (main pic). Science Photo Library: 3br, 10tl (NOAA), 10/11t (Cape Grim BAPS/Simon Fraser), 29bl (Earth Satellite Corp). Spectrum Colour Library: 30cb, 31cr, 30/31b. Still Pictures: 12bl, 13cb, 26/27b. Telegraph Colour Library: 4/5 (main pic), 6/7t, 28/29 (main pic). Tony Stone: 2tl, 2/3 (main pic) & 32, 6/7t, 7b, 15br.

Every effort has been made to trace the copyright holders and we apologize in advance for any unintentional omissions.
We would be pleased to insert the appropriate acknowledgment in any subsequent edition of this publication.

BARRON'S